0 1 2 3 4 5 6 7 8 9 ΔΠΡ 0

1   1
2   2
3   3
4   4
5   5
6   6
7   7
8   8
9   9

FASCINATING FACTS OF DECIMAL DIGITS 0123456789 0
FOR FUN
AMAZEMENT ASTONISHMENT AMUSEMENT

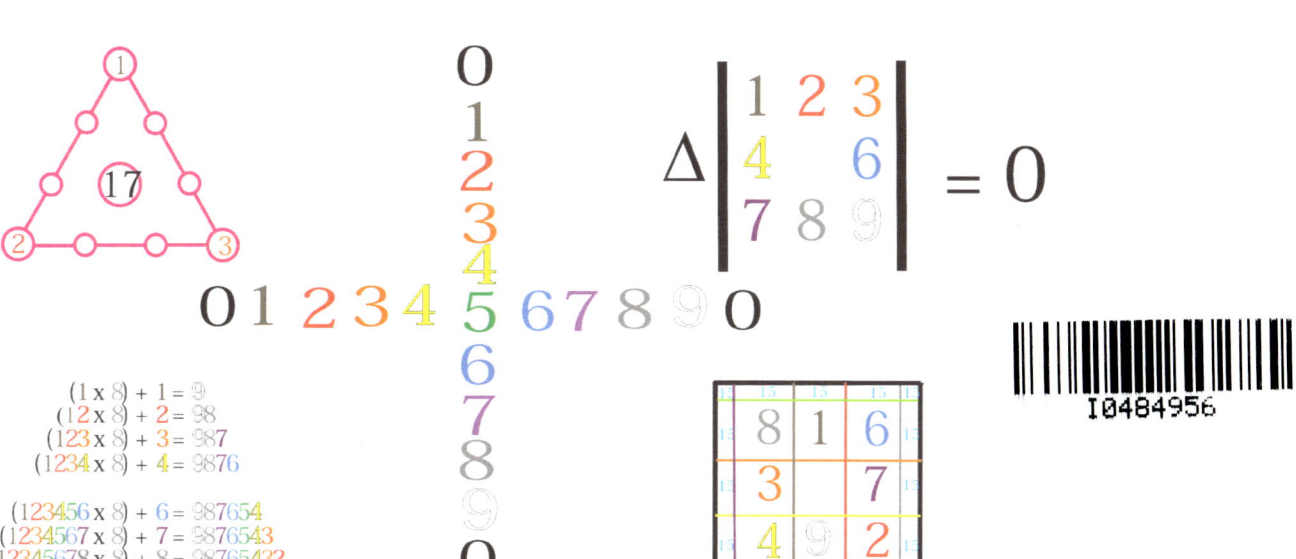

$$\Delta \begin{vmatrix} 1 & 2 & 3 \\ 4 & & 6 \\ 7 & 8 & 9 \end{vmatrix} = 0$$

0 1 2 3 4 5 6 7 8 9 0

(1 x 8) + 1 = 9
(12 x 8) + 2 = 98
(123 x 8) + 3 = 987
(1234 x 8) + 4 = 9876

(123456 x 8) + 6 = 987654
(1234567 x 8) + 7 = 9876543
(12345678 x 8) + 8 = 98765432
(123456789 x 8) + 9 = 987654321

8	1	6
3		7
4	9	2

I0484956

DAVY PETER RAJANAYAGAM

0 1 2 3 4 5 6 7 8 9 0

0 1 2 3 4 5 6 7 8 9 ΔΠΡ 0

1   1
2   2
3   3

To order additional copies of this book, contact:
Xlibris Corporation
0-800-644-6988
www.xlibrispublishing.co.uk
Orders@xlibrispublishing.co.uk

4   4
5   5
6   6
7   7
8   8
9   9

0 1 2 3 4 5 6 7 8 9 0

0 1 2 3 4 5 6 7 8 9 ΔΠΡ 0

FASCINATING FACTS OF DECIMAL DIGITS 01234567890

ACKNOWLEDGMENT

The author wishes to confirm the fact that most of the facts presented have been compiled from the facts that were common knowledge from several published sources over many years,
However the author wants to stress the fact that he has modified and enhanced the facts by extending
(as far as is practical) to cover all the digits 0 to 9 and in one or two cases included 10 even though
it is a two digit figure
The author has color coded the numbers using BSEN 60062 formerly BS 1852 normally used in electronics
The author wants to emphasize the fact that ORIGINAL ideas (may be in some cases independent rediscoveries)
have been incorporated.
For example the page "Some bother to get a palindrome" is the author's own work.
The Trivia attached to the Magic Square is the author's own hard work.
The interesting patterns have all been rediscovered in some cases and modified and extended to include
all the digits 0 to 9.
The Magic Triangle has been developed to cover the digits 4,5&6 and 7,8&9 in the
corners (apex)using cyclic symmetry and other minor changes.
The determinants of digits 0 to 9 is unusual that it is presented as a property of the decimal digits.
The puzzle was a common vulgar joke and was transformed to the current one after extensive trial and error
to get the magic palindrome from the subtraction facts.
In short, even though the material like the multiplication tables have been in circulation for ages
the author after much effort has modified extended and added original ideas to submit them in the
current format in color using BSEN 60062 /BS1852 as mentioned earlier.

0 1 2 3 4 5 6 7 8 9 _{ΔΠΡ} 0

FASCINATING FACTS OF DECIMAL DIGITS 01234567890
FOR FUN
AMAZEMENT ASTONISHMENT AMUSEMENT

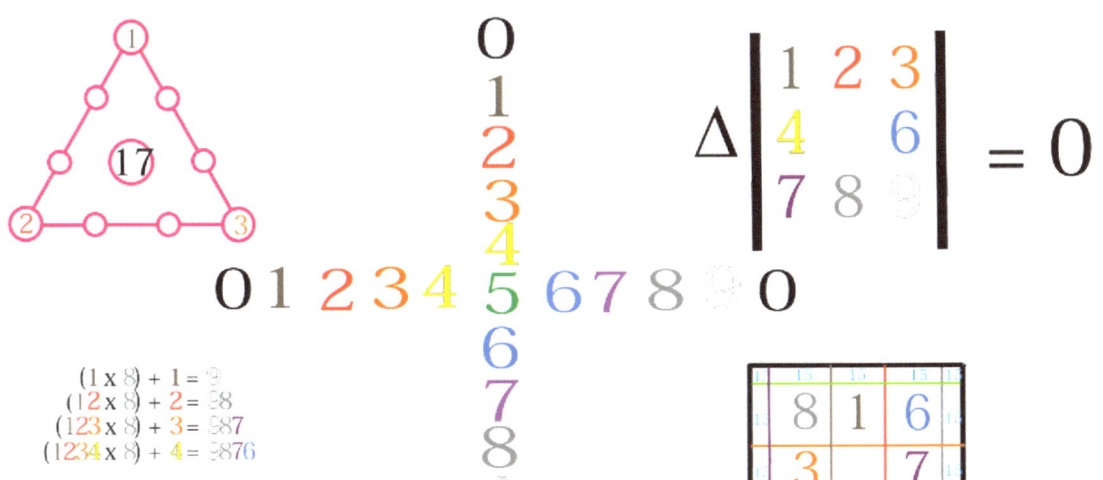

$$\Delta \begin{vmatrix} 1 & 2 & 3 \\ 4 & & 6 \\ 7 & 8 & \end{vmatrix} = 0$$

0 1 2 3 4 5 6 7 8 9 0

(1 x 8) + 1 = 9
(12 x 8) + 2 = 98
(123 x 8) + 3 = 987
(1234 x 8) + 4 = 9876

(123456 x 8) + 6 = 987654
(1234567 x 8) + 7 = 9876543
(12345678 x 8) + 8 = 98765432
(123456789 x 8) + 9 = 987654321

8	1	6
3		7
4		2

DAVY PETER RAJANAYAGAM

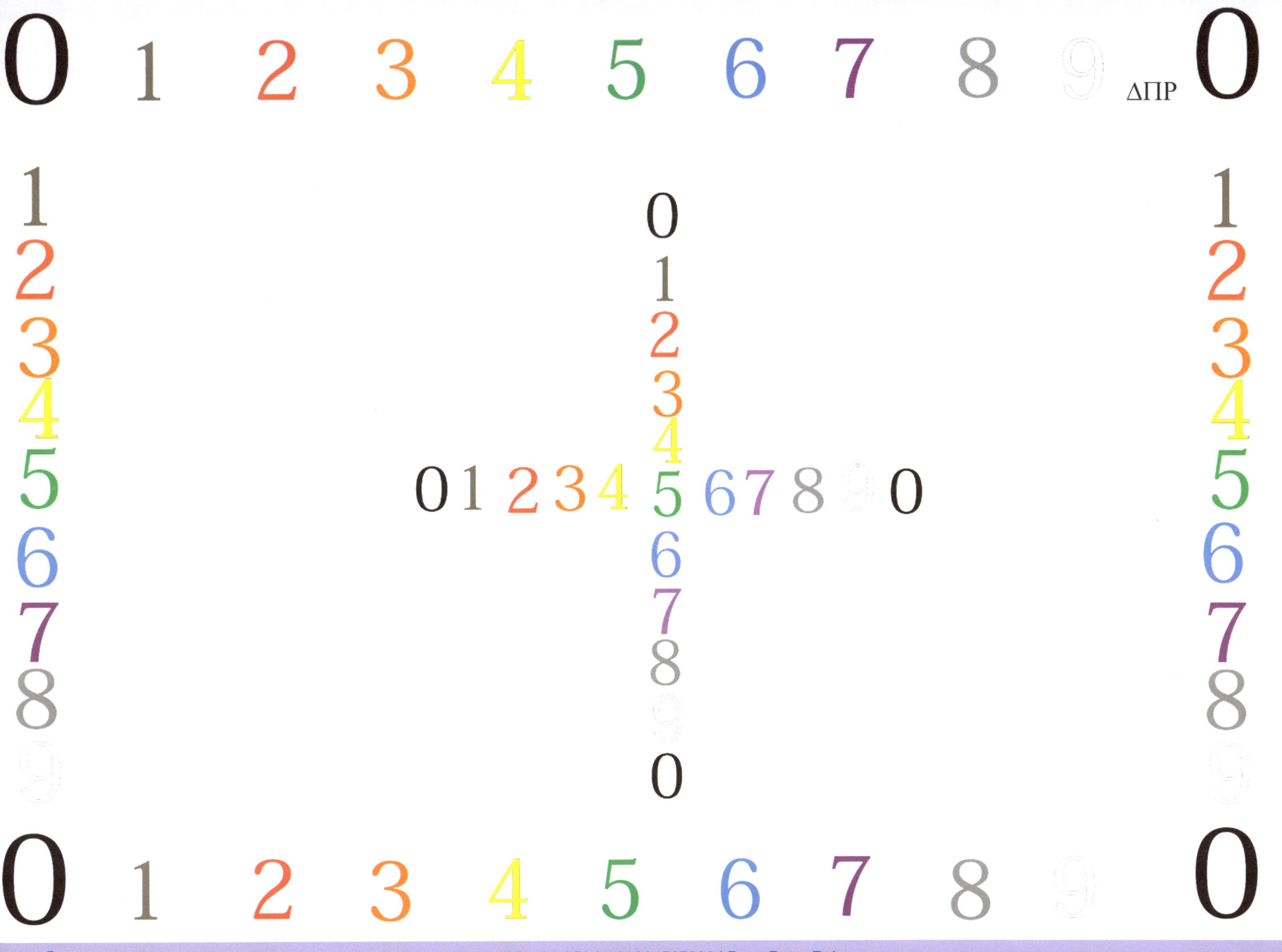

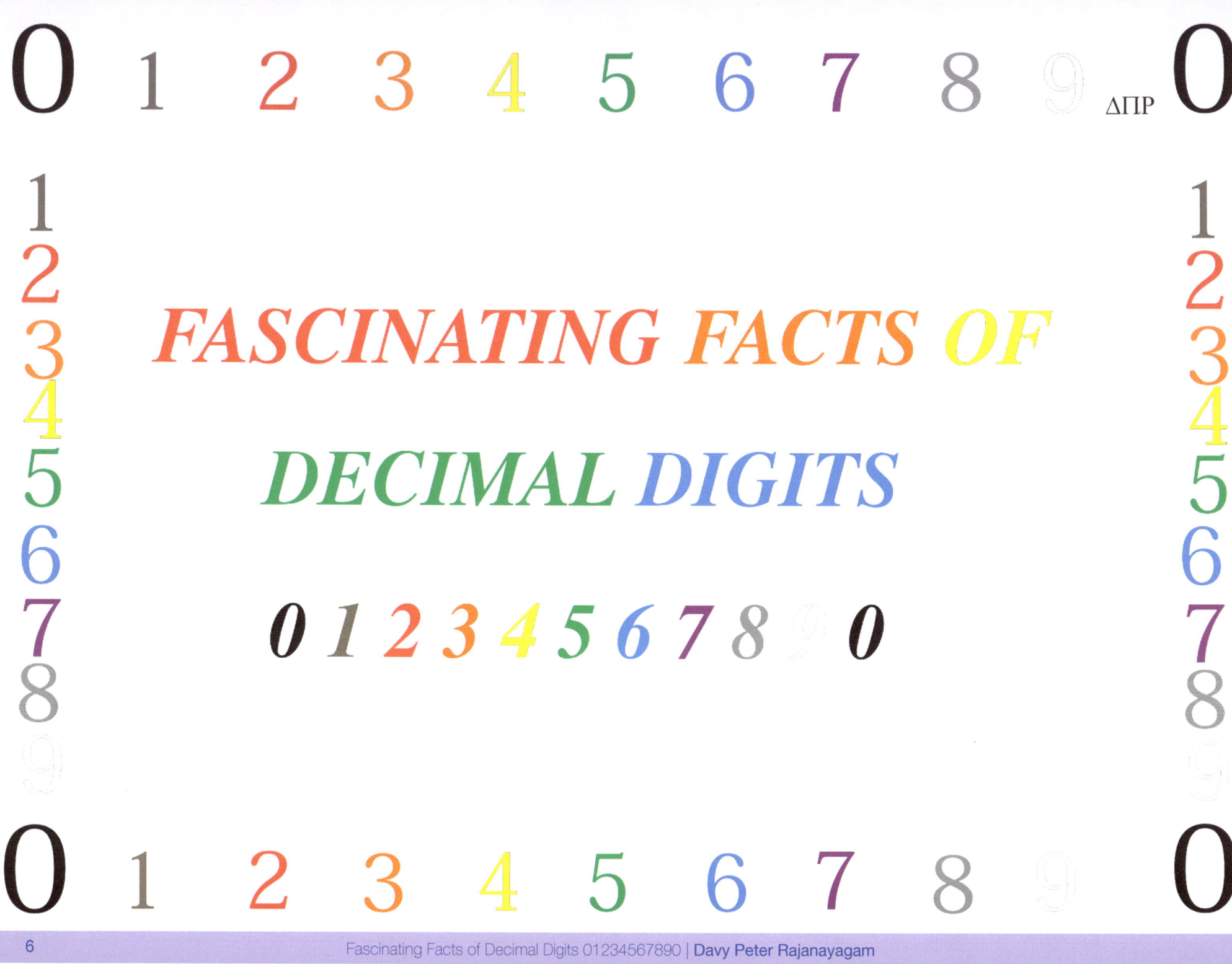

0 1 2 3 4 5 6 7 8 9 ΔΠΡ 0

FASCINATING FACTS OF DECIMAL DIGITS 01234567890

COMPILED - MODIFIED-ORIGINATED

AS THE CASE MAY BE

BY

DAVY PETER RAJANAYAGAM

[{(MEng BSc(Hons) BA)(Open UK)} {(BT BSc)}]

LCGI AMIMA MInstP MIET (M0RVT)

EX. MCollP MIEE (M6RPD) (2E0TAV)

0 1 2 3 4 5 6 7 8 9 0

FASCINATING FACTS OF DECIMAL DIGITS 01234567890

AN INTERESTING PATTERN 1

" α " all ones squared

Symmetric about

DECIMAL DIGITS

Palindromes

$$1^2 = 1$$
$$11^2 = 121$$
$$111^2 = 12321$$
$$1111^2 = 1234321$$
$$11111^2 = 123454321$$
$$111111^2 = 12345654321$$
$$1111111^2 = 1234567654321$$
$$11111111^2 = 123456787654321$$
$$111111111^2 = 12345678987654321$$

$$(1..-.9-...1)$$

FASCINATING FACTS OF DECIMAL DIGITS 01234567890
AN INTERESTING PATTERN 2

"β" Times nine descending order   All Eights

$$(0 \times 9) + 8 = 8$$
$$(9 \times 9) + 7 = 88$$
$$(98 \times 9) + 6 = 888$$
$$(987 \times 9) + 5 = 8888$$
$$(9876 \times 9) + 4 = 88888$$
$$(98765 \times 9) + 3 = 888888$$
$$(987654 \times 9) + 2 = 8888888$$
$$(9876543 \times 9) + 1 = 88888888$$
$$(98765432 \times 9) + 0 = 888888888$$
$$(987654321 \times 9) - 1 = 8888888888$$

0 1 2 3 4 5 6 7 8 9 ΔΠΡ 0

FASCINATING FACTS OF DECIMAL DIGITS 012345678 0
AN INTERESTING PATTERN 3

"γ"Times nine ascending order

All Ones

$$(0 \times 9) + 1 = 1$$
$$(1 \times 9) + 2 = 11$$
$$(12 \times 9) + 3 = 111$$
$$(123 \times 9) + 4 = 1111$$
$$(1234 \times 9) + 5 = 11111$$
$$(12345 \times 9) + 6 = 111111$$
$$(123456 \times 9) + 7 = 1111111$$
$$(1234567 \times 9) + 8 = 11111111$$
$$(12345678 \times 9) + 9 = 111111111$$
$$(123456789 \times 9) + 10 = 1111111111$$

0 1 2 3 4 5 6 7 8 9 ΔΠΡ 0

FASCINATING FACTS OF DECIMAL DIGITS 01234567890
AN INTERESTING PATTERN 4

" δ "Times eight ascending order   Reverse Digits

$$(1 \times 8) + 1 = 9$$
$$(12 \times 8) + 2 = 98$$
$$(123 \times 8) + 3 = 987$$
$$(1234 \times 8) + 4 = 9876$$
$$(12345 \times 8) + 5 = 98765$$
$$(123456 \times 8) + 6 = 987654$$
$$(1234567 \times 8) + 7 = 9876543$$
$$(12345678 \times 8) + 8 = 98765432$$
$$(123456789 \times 8) + 9 = 987654321$$

0 1 2 3 4 5 6 7 8 9 ΔΠΡ 0
1 1

FASCINATING FACTS OF DECIMAL DIGITS 012345678 0
AN INTERESTING PATTERN 5

" ε " Times nine digits ascending order right to left

Digits followed by Eights

$$(1 \text{ x } 9) - 1 = 08$$
$$(21 \text{ x } 9) - 1 = 188$$
$$(321 \text{ x } 9) - 1 = 2888$$
$$(4321 \text{ x } 9) - 1 = 38888$$
$$(54321 \text{ x } 9) - 1 = 488888$$
$$(654321 \text{ x } 9) - 1 = 5888888$$
$$(7654321 \text{ x } 9) - 1 = 68888888$$
$$(87654321 \text{ x } 9) - 1 = 788888888$$
$$(987654321 \text{ x } 9) - 1 = 8888888888$$
$$*(10987654321 \text{ x } 9) - 1 = 98888888888$$

FASCINATING FACTS OF DECIMAL DIGITS 01234567890

AN INTERESTING PATTERN 6

Merging "α" & "γ"

Gives this new pattern

$$\{(0 \times 9) + 1\}^2 = 1$$
$$\{(1 \times 9) + 2\}^2 = 121$$
$$\{(12 \times 9) + 3\}^2 = 12321$$
$$\{(123 \times 9) + 4\}^2 = 1234321$$
$$\{(1234 \times 9) + 5\}^2 = 123454321$$
$$\{(12345 \times 9) + 6\}^2 = 12345654321$$
$$\{(123456 \times 9) + 7\}^2 = 1234567654321$$
$$\{(1234567 \times 9) + 8\}^2 = 123456787654321$$
$$\{(12345678 \times 9) + 9\}^2 = 12345678987654321$$

FASCINATING FACTS OF DECIMAL DIGITS 01234567890

THE MAGIC TRIANGLE 1

CORNERS 1 2 3 Version 1 Anti clockwise

SUM ALONG SIDES 17

4 5 6 & 7 8 9 CYCLIC CLOCKWISE
STARTING AT BASE AND LHS

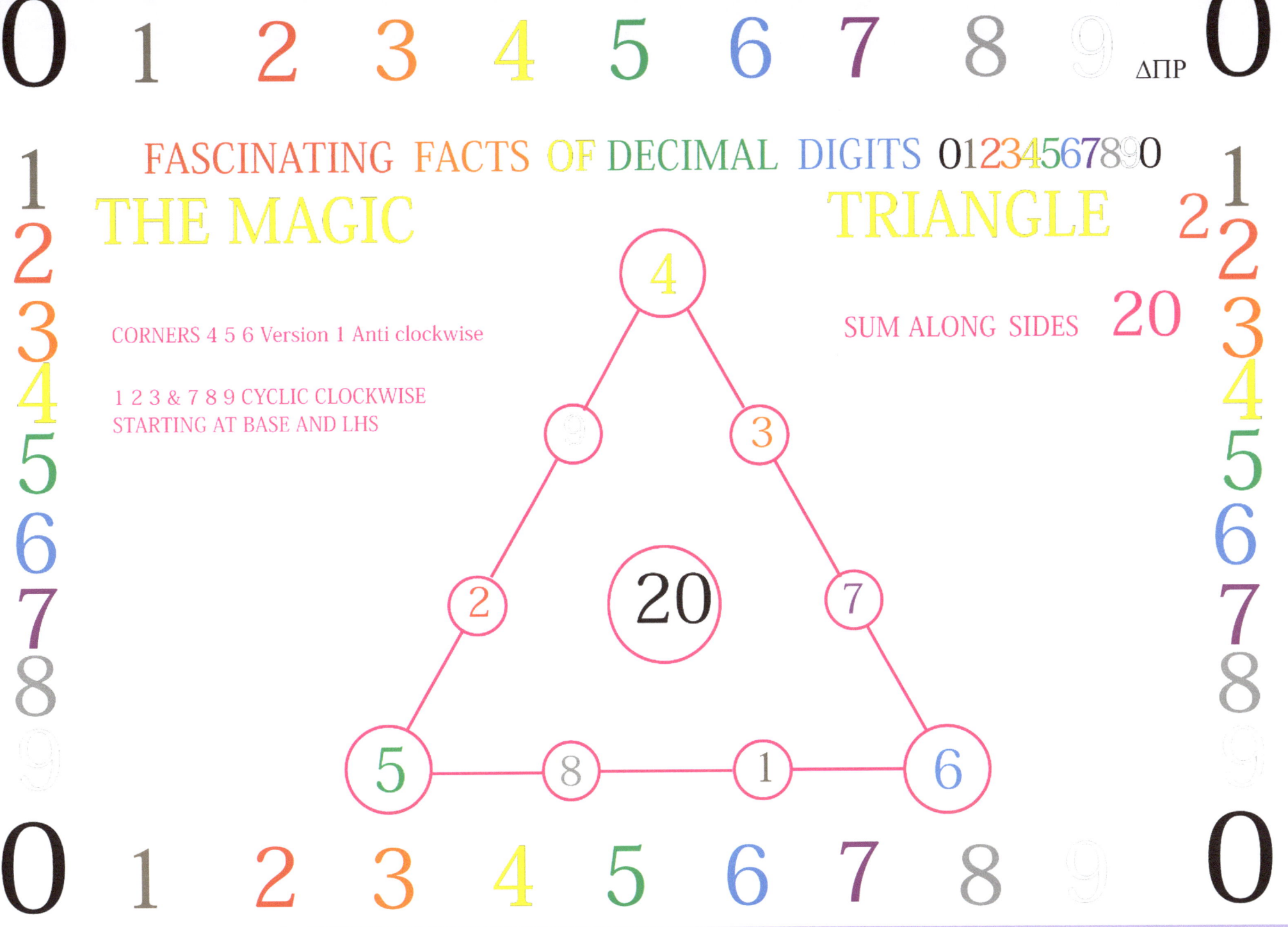

FASCINATING FACTS OF DECIMAL DIGITS 01234567890

THE MAGIC   TRIANGLE

CORNERS 4 5 6 Version 1 Anti clockwise   SUM ALONG SIDES 20

1 2 3 & 7 8 9 CYCLIC CLOCKWISE
STARTING AT BASE AND LHS

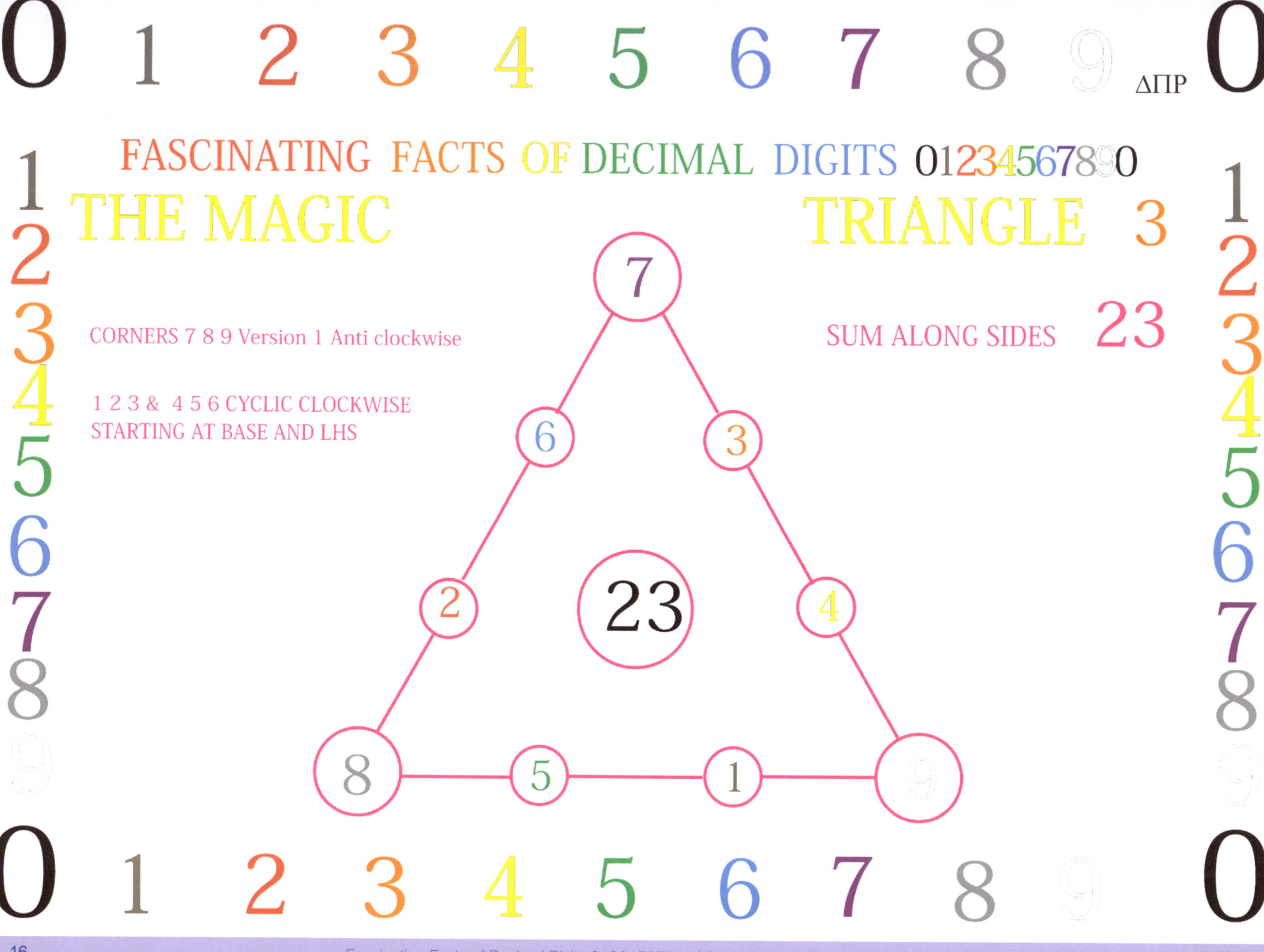

FASCINATING FACTS OF DECIMAL DIGITS 01234567890

THE MAGIC   TRIANGLE 3

CORNERS 7 8 9 Version 1 Anti clockwise   SUM ALONG SIDES   23

1 2 3 & 4 5 6 CYCLIC CLOCKWISE
STARTING AT BASE AND LHS

FASCINATING FACTS OF DECIMAL DIGITS 01234567890

THE MAGIC   TRIANGLE 4

CORNERS 1 2 3 Version 2 Anti clockwise

4 5 6 & 7 8 9 CYCLIC CLOCKWISE
does not apply

SUM ALONG SIDES 17

Sum along sides of TOP Triangle also 17

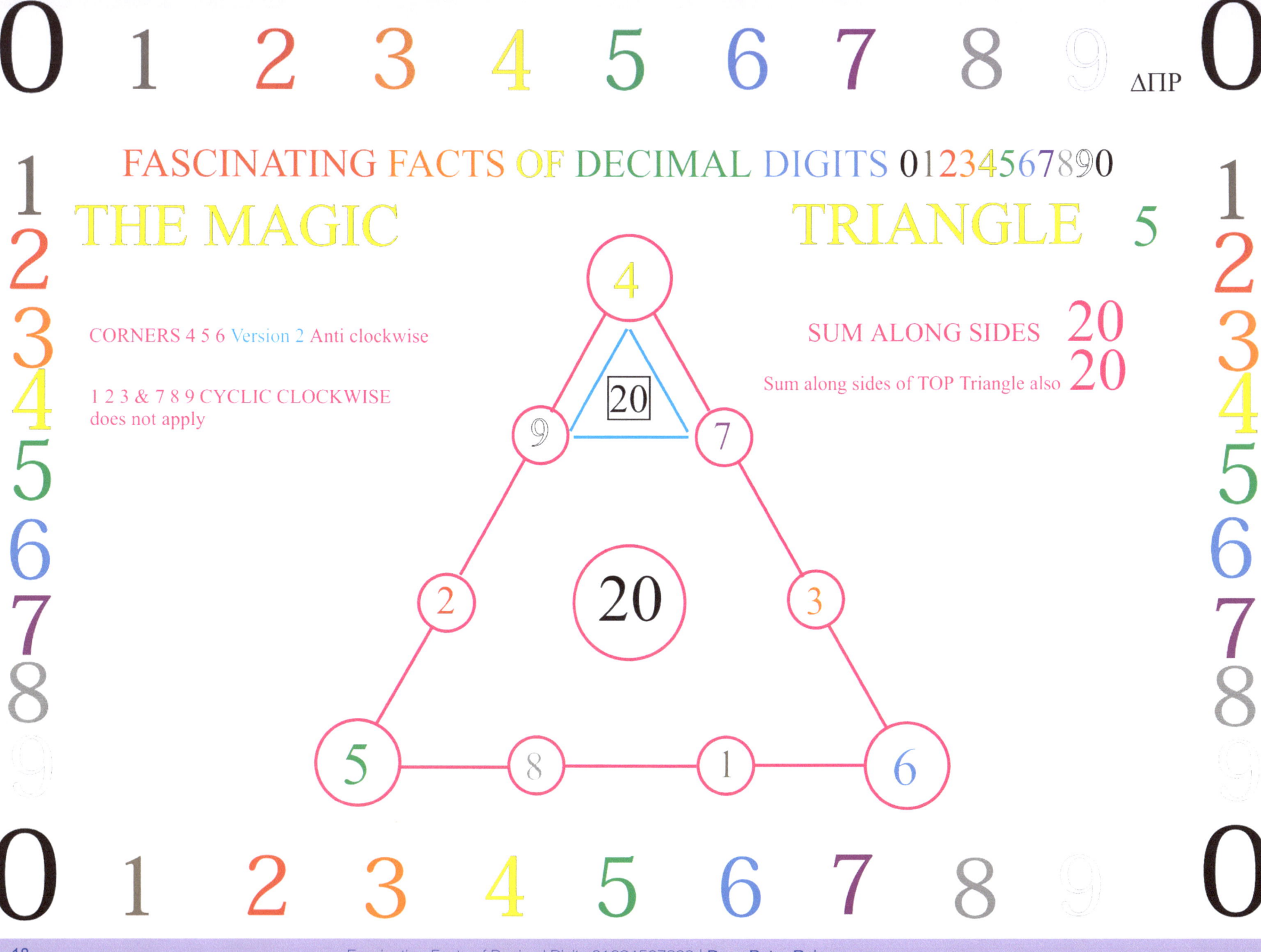

FASCINATING FACTS OF DECIMAL DIGITS 01234567890

THE MAGIC   TRIANGLE

CORNERS 7 8 9 Version2 Anti clockwise

SUM ALONG SIDES   23

THE NUMBERS ON EITHER SIDE OF 23 ARE 2 & 3

1 2 3 & 4 5 6 CYCLIC CLOCKWISE
does not apply

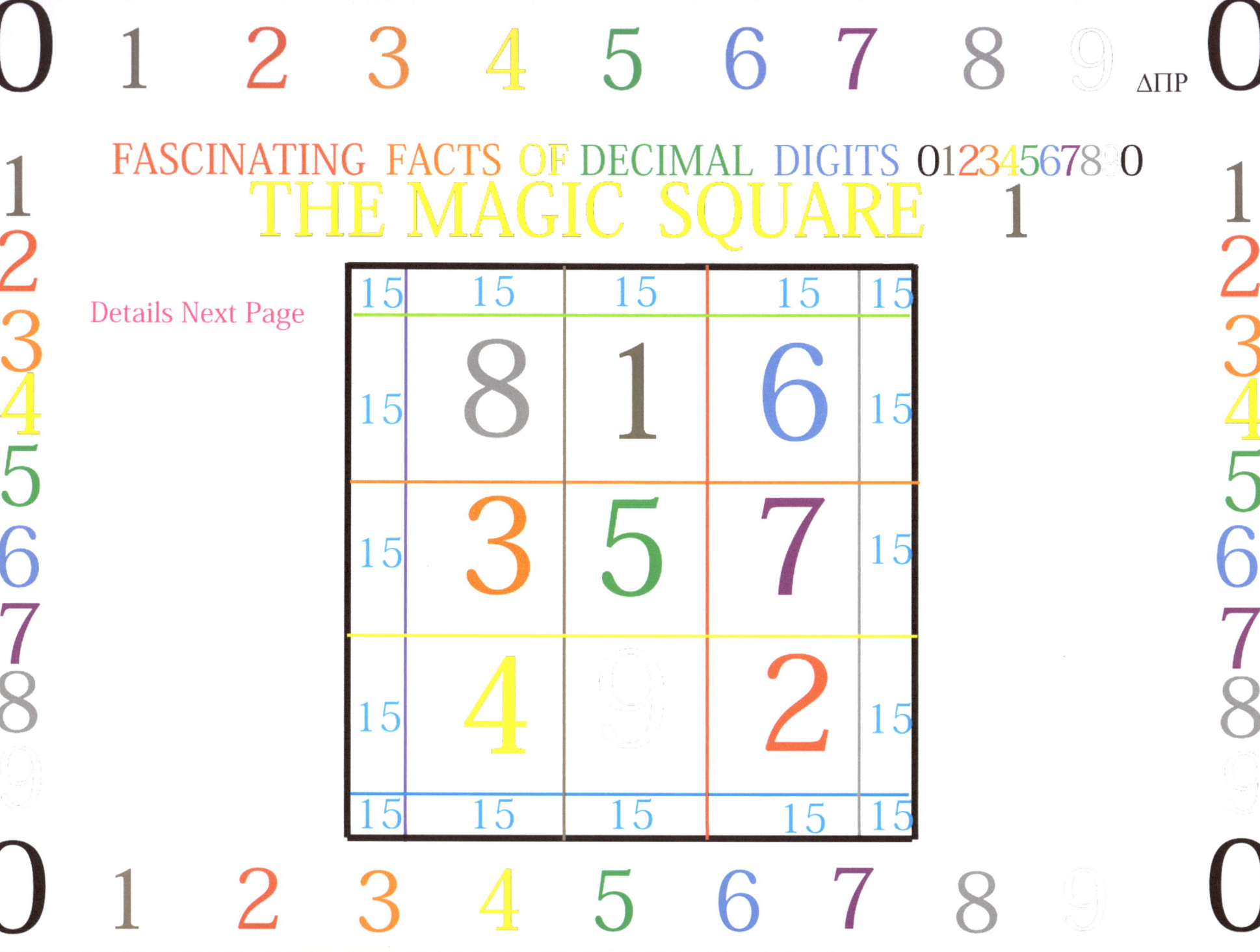

FASCINATING FACTS OF DECIMAL DIGITS 0123456789 0

THE MAGIC SQUARE

Details Next Page

THE MAGIC SQUARE 2

TRIVIA

The magic square is formed by placing 5 in the centre square. Then the Even numbers in the corner squares diagonally opposite to each other. The numbers being Complements of 10. The Odd numbers occupy the inner squares and are also Complements of 10.

The SUM along the rows, along the columns and along the diagonals add to the Magic sum of 15
Let the sum S = 15
There are 16 positions of 15

$$S = 15$$

$$\Delta = (16 \times S^2)/10 = 360 = (S^2 + S^3)/10 = \Delta$$

Determinant $\Delta \begin{vmatrix} 8 & 1 & 6 \\ 3 & 5 & 7 \\ 4 & 9 & 2 \end{vmatrix} = 360$

FASCINATING FACTS OF DECIMAL DIGITS 01234567890
DETERMINANTS OF DECIMAL DIGITS ONE TO NINE

$$\Delta \begin{vmatrix} 1 & 2 & 3 \\ 4 & 5 & 6 \\ 7 & 8 & 9 \end{vmatrix} = 0 \qquad \Delta \begin{vmatrix} 9 & 8 & 7 \\ 6 & 5 & 4 \\ 3 & 2 & 1 \end{vmatrix} = 0$$

PLEASE SEE
NEXT PAGE

$$\Delta \begin{vmatrix} 1 & 4 & 7 \\ 2 & 5 & 8 \\ 3 & 6 & 9 \end{vmatrix} = 0 \qquad \Delta \begin{vmatrix} 9 & 6 & 3 \\ 8 & 5 & 2 \\ 7 & 4 & 1 \end{vmatrix} = 0$$

FASCINATING FACTS OF DECIMAL DIGITS 01234567890
DETERMINANTS OF DECIMAL DIGITS ONE TO NINE

$$\Delta \begin{vmatrix} 1 & 2 & 3 \\ 4 & 5 & 6 \\ 7 & 8 & 9 \end{vmatrix} = 0$$

**TO FIND DETERMINANT PROCEED AS FOLLOWS
WRITE THE DETERMINANT AS SHOWN BELOW**

The determinants written in the special way in Rows and Columns as above have the property that their determinants are the same

$$\begin{vmatrix} 1 & 2 & 3 \\ 4 & 5 & 6 \\ 7 & 8 & 9 \end{vmatrix} \qquad \begin{vmatrix} 1 & \rule{1cm}{0.4pt} \\ & 5 & 6 \\ & 8 & 9 \end{vmatrix} \qquad \begin{vmatrix} \rule{0.5cm}{0.4pt} & 2 & \rule{0.5cm}{0.4pt} \\ 4 & & 6 \\ 7 & & 9 \end{vmatrix} \qquad \begin{vmatrix} \rule{1cm}{0.4pt} & 3 \\ 4 & 5 \\ 7 & 8 \end{vmatrix}$$

Then write as below

1[9x5 - 8x6] - 2[9x4 - 7x6] + 3[8x4 - 7x5]

=1[45 - 48] -2[36 - 42] + 3[32 - 35]

=1x(-3) -2(-6)+3(-3)= (-3)+12-(9)=0

Use similar methods for the other determinants specially for the Magic Square

0 1 2 3 4 5 6 7 8 9 ΔΠΡ 0

FASCINATING FACTS OF DECIMAL DIGITS 01234567890
CHAOS TO CONSOLATION THROUGH CALCULATION
A PUZZLE USING A PROPERTY OF DECIMAL DIGITS 0 to 9

9 8 7 6 5 4 3 2 1 0

1. From the above list Choose ANY THREE CONSECUTIVE NUMBERS IN DESCENDING ORDER and write it down and call it "P"

2. Invert the above number [that is write in Reverse order (Left to Right)(that is in Ascending order) and write it down below "P" and call the new number "Q"

3. SUBTRACT "Q" from "P" Write down the result call it "R"

4. Multiply "R" by the Magic Palindrome "4 3 6 4 6 3 4"

5. Write down the ANSWER to 4 and keep it handy.
SUBSTITUTE the DIGITS by their corresponding LETTERS given in the Clue/s in the next page and read the SOLUTION.

FASCINATING FACTS OF DECIMAL DIGITS 0123456789 0
CHAOS TO CONSOLATION THROUGH CALCULATION
CLUES TO AND SOLUTION OF PUZZLE IN THE PREVIOUS PAGE

SOLUTIONS

CLUE 1.
1 2 3 4 5 6 7 8 9 0
I E V D O O L G S
{1 Jo 4:8 : 4:16}
DAVY PETER

CLUE 2
1 2 3 4 5 6 7 8 9 0
I S A R M H T C S
{Birth Day of JESUS}
MARGARET M

CLUE 3
1 2 3 4 5 6 7 8 9 0
Z H T I E L B E A R
{The Sovereign-UK}
LOVE AND JOY

CLUE 4
1 2 3 4 5 6 7 8 9 0
C E V A O E L P E
{Fruits of the Spirit Gal 5:22}
LOVE PEACE

CLUE 5
1 2 3 4 5 6 7 8 9 0
E O J V D O N L A Y
{Fruits of the Spirit Gal 5:22}
ELIZABETH M

CLUE 6
1 2 3 4 5 6 7 8 9 0
G M T R E A R M A
{The Author's wife}
CHRISTMAS

CLUE 7
1 2 3 4 5 6 7 8 9 0
Y R E V T A E D P
{The Author}
GOD IS LOVE

FASCINATING FACTS OF DECIMAL DIGITS 01234567890

ADDITION   PATTERN

It is easier to do the addition given that the sum of "n" consecutive natural numbers is

$$S = n \times (n+1)/2$$

in this case {n = 9}

S = 45

```
0 1 2 3 4 5 6 7 8 9 0
1 2 3 4 5 6 7 8 9 0 1
2 3 4 5 6 7 8 9 0 1 2
3 4 5 6 7 8 9 0 1 2 3
4 5 6 7 8 9 0 1 2 3 4
5 6 7 8 9 0 1 2 3 4 5
6 7 8 9 0 1 2 3 4 5 6
7 8 9 0 1 2 3 4 5 6 7
8 9 0 1 2 3 4 5 6 7 8
9 0 1 2 3 4 5 6 7 8 9
0 1 2 3 4 5 6 7 8 9 0
                  + 1 0
```

SUM =   5 0 1 2 3 4 5 6 7 8 9 5

Note that the sum is made up of all the digits 0 to 9 flanked by 5

FASCINATING FACTS OF DECIMAL DIGITS 01234567890

ADDITION   FACTS

```
  1  2  3  4  5  6  7  8  9              8  5  9         8  5  3
  9  8  7  6  5  4  3  2  1           +  7  4  3      +  7  4  9
+                          1           ─────────      ─────────
─────────────────────────────          1  6  0  2      1  6  0  2
  1  1  1  1  1  1  1  1  1  1  {ALL ONES}
```

ALL DIGITS (0 to 9) ARE USED

```
  1  2  3  4  5  6  7  8  9  0              8  4  9         8  4  3
  0  9  8  7  6  5  4  3  2  1           +  7  5  3      +  7  5  9
+                          1  1          ─────────      ─────────
─────────────────────────────────         1  6  0  2      1  6  0  2
  2  2  2  2  2  2  2  2  2  2  {ALL TWOS}
```

0 1 2 3 4 5 6 7 8 9 ΔΠΡ 0

FASCINATING FACTS OF DECIMAL DIGITS 01234567890

SUBTRACTION   FACTS

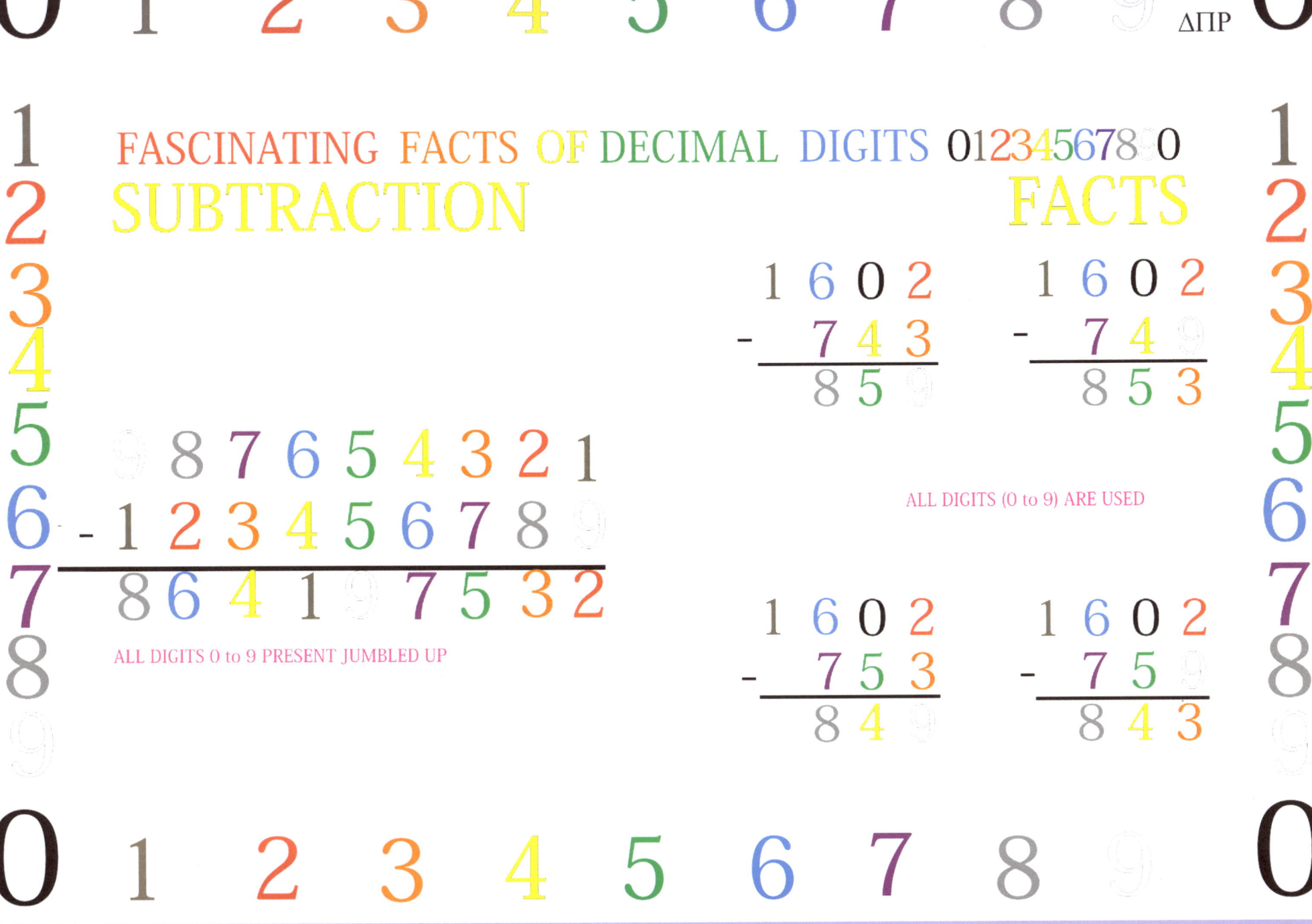

```
  1 6 0 2          1 6 0 2
-     7 4 3      -     7 4 9
  ─────────        ─────────
    8 5 9            8 5 3
```

ALL DIGITS (0 to 9) ARE USED

```
  9 8 7 6 5 4 3 2 1
- 1 2 3 4 5 6 7 8 9
  ─────────────────
  8 6 4 1 9 7 5 3 2
```

ALL DIGITS 0 to 9 PRESENT JUMBLED UP

```
  1 6 0 2          1 6 0 2
-     7 5 3      -     7 5 9
  ─────────        ─────────
    8 4 9            8 4 3
```

FASCINATING FACTS OF DECIMAL DIGITS 01234567890

SOME BOTHER TO GET A PALINDROME USING 123456789 & 10

A. Natural Numbers	1	2	3	4	5	6	7	8	9	10
B. Prime Numbers	1	2	3	5	7	11	13	17	19	23
C. Sum of first "N" Primes	1	3	6	11	18	29	42	59	78	101
D. Square of Natural Numbers	1	4	9	16	25	36	49	64	81	100
E. Difference of "C" & "D"		1	3	5	7	7	7	5	3	1
F. Add Magic Number						2	0	0	0	0
G. Read PALINDROME		1	3	5	7	9	7	5	3	1

All ODD Digits

0 1 2 3 4 5 6 7 8 9 ΔΠΡ 0

FASCINATING FACTS OF DECIMAL DIGITS 0123456789 0
THE TIMES "ONE" & "NINE" TABLES

ASCENDING ORDER	DESCENDING ORDER
0 x 1 = 0 0	9 0 = x 10
1 x 1 = 0 1	8 1 = x 0
2 x 1 = 0 2	7 2 = x 08
3 x 1 = 0 3	6 3 = x 07
4 x 1 = 0 4	5 4 = x 06
5 x 1 = 0 5	4 5 = x 05
6 x 1 = 0 6	3 6 = x 04
7 x 1 = 0 7	2 7 = x 03
8 x 1 = 0 8	1 8 = x 02
9 x 1 = 0	0 9 = x 01

Note that all the digits 0 -9 are present in the Units column and also the 10's column in nines table (Ascending and Descending) .. These digits will be in the reverse order when both tables are in ascending order. As a matter of fact these reversals occur in tables that are Complements of 10 (1&9 3&7 2&8 4&6). Note also in the Results columns the "tens" and "units" digits add upto 9 in each row

0 1 2 3 4 5 6 7 8 9 0

0 1 2 3 4 5 6 7 8 9 ΔΠΡ 0

FASCINATING FACTS OF DECIMAL DIGITS 01234567890

THE TIMES "THREE" & "SEVEN" TABLES

ASCENDING ORDER	DESCENDING ORDER
1 x 3 = 0 3	6 3 = 7 x 9
2 x 3 = 0 6	5 6 = 7 x 8
3 x 3 = 0 9	4 9 = 7 x 7
4 x 3 = 1 2	4 2 = 7 x 6
5 x 3 = 1 5	3 5 = 7 x 5
6 x 3 = 1 8	2 8 = 7 x 4
7 x 3 = 2 1	2 1 = 7 x 3
8 x 3 = 2 4	1 4 = 7 x 2
9 x 3 = 2 7	0 7 = 7 x 1

Note that all the digits 1-9 are present in the Units column though jumbled up.(but in the same order). These digits will be in the reverse order when both tables are in ascending order. As a matter of fact these reversals occur in tables that are Complements of 10 (1&9 3&7 2&8 4&6).

0 1 2 3 4 5 6 7 8 9 0

0 1 2 3 4 5 6 7 8 9 ΔΠΡ 0

1
2
3
4
5
6
7
8
9

0 1 2 3 4 5 6 7 8 9 0

1
2
3
4
5
6
7
8
9